ISBN 978-1-332-61639-8
PIBN 10260706

O

METEOROLOGICAL TABLES

AND

CLIMATOLOGY OF VERMONT,

WITH MAP SHOWING THE RAINFALL;

ALSO,

SUGGESTIONS AND DIRECTIONS ABOUT FORETELLING STORMS.

Adolphus

By HIRAM A. CUTTING, A. M., M. D.,

STATE GEOLOGIST AND CURATOR.

²MONTPELIER:

J. & J. M. POLAND, OFFICIAL STATE PRINTERS.

CLIMATOLOGY OF VERMONT.

Observations made in the interests of meteorology, and collected and reduced under the patronage of our government, have shown that though our storms are irregular, that our hills and valleys are rendered fertile by the moisture brought from the Gulf of Mexico, by the great south west current. This is modified by the physical contour of the country, and by the great atmospheric current coming from the Pacific, which, though it loses most of its moisture before it reaches us, it modifies our climate, as we get westerly winds at least three fourths of the time ; and sometimes when we get falling weather, this wind, meeting the moisture coming inland from the Atlantic, condenses it, causing showers, and now and then extensive storms. Thus the atmospheric disturbances generated thousands of miles away, give us in connection with our local storms sufficient moisture to render our State fertile, even to the mountain summits.

Through the investigations made by Prof. Joseph Henry, of the Smithsonian Institute, we see that our great north east storms have their origin within the tropics. The great current of the south east trade wind infringes on the north east trade wind, and produces the cyclones so destructive on the coast of Florida and the West Indies. The cyclone thus generated moves along the coast with diminished violence, producing our north east storms. Sometimes this disturbance originates among the Rocky Mountains, sweeping over the continent. This takes place oftener in August or midsummer ; while in autumn and winter we are more dependent upon the Gulf storms. From observations of the Signal Service, it is shown that from seven to thirteen areas of low barometer are developed per month within or upon the borders of the United States. From one to three of those pass directly over Vermont. As a rule rainfall increases as you go towards the equator, and decreases towards the poles ; thus we should expect more rainfall in the southern part of the State, and such is true, yet its geographical distribution depends somewhat upon the contour of our State, as the map shows. Though they may receive more rain, it is in no way certain that they may not more often suffer from drought than other sections, as the nature of the soil and the greater or less humidity of the atmosphere from surrounding forests or ponds may much modify evaporation. The amount of vapor in the atmosphere regulates the intensity of the sun's rays, as well as preventing the radiation of the sun's heat from the earth, which radiation rapidly aids evaporation of moisture. Thus in some sections crops may severely suffer from drought, while in others they may not, yet the rainfall for the time may be equal. Thus it will be seen that the preservation of vegetation upon our hilltops and mountains not only modifies the effects of drought, but, carrying out the same reasoning, moderates the extremes of winter. In relation to soil, it is easy to ascertain which will retain the most moisture, by taking a given amount that is perfectly dry, and see how much water you can add to it without its leaching, and then how rapidly it loses it by evaporation. These experiments show that while the usual soil of our hillsides is losing three fourths that of our valleys, which will not at first retain as much, will lose nine tenths of what it at first absorbed, thus showing why our hills are always green, and hill land more sure not to suffer from drought.

The foregoing map shows, not only the annual aqueous precipitation for the State, but for a large part of New Hampshire, and a part of Massachusetts and New York. The greatest amount of annual rainfall seems to be about the highest of the White Mountains. It is set down at fifty-five inches, but the rainfall for 1876 was seventy-eight inches. It is probable that fifty-five inches is none too high for the average. The next largest is in central New Hampshire and southern Vermont, being from forty-four to forty-six inches, shading northward in Vermont to thirty-eight, thirty-nine, and forty inches, while an area west of Lake Champlain has but thirty-seven inches.

In comparison with other places, this seems a small amount. The average annual rainfall in Calcutta is 118 inches. On the Amazon river, 500 inches. In Guadeloupe, 282 inches. In Singapore, 190 inches, and many other places equally large, but the general summary of observations throughout the world establishes the fact that within the tropics the average amount is about 100 inches; and the average number of rainy days eighty. In a belt, including the Southern States, seventy inches in one hundred and twenty-five rainy days. In a belt from the Southern States to northern Vermont, forty-five inches in one hundred and fifty-six stormy days, while the average fall in Siberia is only five and a half inches in two hundred stormy days. A small amount of rain if at proper intervals is sufficient for vegetation. Our amount of rainfall is as well calculated for production as that of any section of the globe. It much exceeds many sections of the fertile West, and is much more constant from year to year.

The following table shows the amount of annual, as well as monthly, rainfall at forty-three stations, most of them being for a series of years:

TABLE OF RAINFALL.

COMPILED FROM MANUSCRIPT RECORDS, SMITHSONIAN TABLES, AND OTHER AUTHENTIC SOURCES.

NAMES OF STATIONS.	Jan.	Feb.	March.	April	May	June	July	Aug.	Sept.	Oct.	Nov.	Dec.	Spring	Sumr.	Aut.	Win.	Annual Rain-fall.	REMARKS.
Barnet, Vt.....	1.86	3.86	3.20	2.06	4.70	4.75	4.50	3.50	2.50	6.50	2.75	1.00	9.96	12.76	11.75	6.70	41.16	Compiled from observations of 1869.
Bennington, "	2.64	2.38	2.62	2.68	3.32	2.79	4.54	4.66	3.18	3.18	3.00	2.18	8.47	11.99	9.36	7.40	44.00	Monthly means could not be obtained.
Brandon, "	1.96	2.19	3.05	2.69	2.46	2.84	4.18	3.51	4.71	3.76	2.72	2.41	10.10	10.58	11.14	5.86	37.22	From years 1856 to 1869, from 1870 to 1874, give 38.29 in.
Burlington, "	1.87	1.63	2.00	2.47	2.43	2.51	2.51	2.64	2.64	5.67	1.99	2.84	6.90	12.15	10.30	6.34	37.63	Smithsonian tables give a little less rainfall.
Castleton, "	2.80	2.40	2.00	2.67	3.45	4.12	2.95	4.69	4.40	4.21	3.10	2.85	8.12	12.75	11.71	7.55	35.69	Earlier observations give 38.44.
Craftsbury, "	2.90	2.71	2.71	4.29	4.70	4.66	5.00	4.12	3.37	3.18	3.67	2.60	11.74	13.78	10.22	8.21	40.13	From 1867 to 1873. Smithsonian tables give a little more rain.
Fayetteville, "	1.57	2.86	3.04	4.26	2.50	3.40	4.70	4.04	3.25	4.46	1.98	2.56	9.90	13.14	9.69	6.49	43.95	From 1873 and 1874.
Ferrisburg, "	2.30	2.71	2.42	4.06	2.70	4.11	4.00	4.02	3.25	3.71	2.98	2.30	9.18	12.18	10.59	7.88	38.12	Smithsonian tables show a little more rain.
Hyde-Park, "	3.32	2.09	3.17	3.09	4.27	2.59	3.48	3.87	4.12	3.94	3.37	4.15	10.53	9.57	11.33	9.46	39.28	From 1874 and 1875.
Lunenburgh, "	2.19	2.81	3.46	2.86	3.66	2.96	3.22	3.39	4.41	3.90	3.37	2.40	8.98	9.56	11.37	7.40	41.19	From a series of twenty-nine years, 1849 to 1877.
Middlebury, "	3.30	3.00	2.85	2.75	4.18	2.68	3.70	4.15	3.92	2.70	2.50	3.70	9.78	10.53	9.12	9.90	57.21	From 1864 to 1870. Average since that date, 37.96.
Montpelier, "	2.80	2.76	2.80	3.43	3.43	3.50	3.50	3.80	3.40	4.00	2.40	2.47	9.35	11.10	10.65	8.94	39.18	No monthly means attainable.
Newport, "	2.78	2.91	2.75	3.28	3.57	3.41	3.36	3.90	3.40	4.08	2.31	2.47	8.40	10.68	11.29	8.16	39.33	From March, 1871 to 1877, also 1875 to 1877.
Norwich, "	2.50	2.78	3.10	3.01	4.72	2.91	2.81	2.38	3.01	3.68	4.10	3.49	10.83	8.60	10.74	9.77	40.04	From 1873 to 1876, also Smithsonian tables.
Randolph, "	2.17	1.46	2.68	2.96	3.09	4.08	4.88	3.55	3.59	5.01	3.90	3.69	8.73	12.46	12.50	7.32	28.51	Reduced from eight years' observations.
Rutland, "	4.54	2.57	3.40	3.59	3.58	4.08	5.16	4.13	2.70	3.94	3.02	2.75	10.52	13.31	9.66	9.66	28.94	For the year 1769. 1875 gives 41.20.
St. Johnsb'y, "	2.21	2.08	3.48	3.69	2.37	2.65	5.86	4.38	3.64	3.84	3.02	3.68	9.43	12.83	9.11	7.97	41.01	From 1859 to 1882.
Stowe, "	2.63	2.76	3.02	3.19	3.35	4.29	5.49	4.54	4.57	5.10	3.67	3.65	9.74	14.33	12.41	7.38	29.24	Could not get monthly rainfall.
Springfield, "	2.90	2.44	2.62	3.78	3.86	4.60	4.34	2.96	4.57	3.40	2.17	2.08	11.12	11.88	10.14	7.70	44.15	From 1860 to 1864.
Troy,	3.11	2.77	3.70	3.88	3.54	3.18	4.48	4.29	3.77	3.65	2.76	3.08	9.10	10.35	11.52	8.91	39.44	Combined with observations at South Troy.
W. Charlotte, "	4.02	2.56	4.28	3.41	3.39	2.50	4.82	4.18	3.51	5.54	4.57	3.57	10.47	11.00	13.88	11.15	43.30	Seems too high for verification of stations near. Mean of six years.
Windsor, "	3.84	3.56	4.98	4.28	3.08	5.02	5.12	3.50	3.61	5.51	3.84	3.65	10.82	11.64	12.90	10.72	40.85	Compiled from different observers.
Woodstock, "	2.96	2.26	4.99	4.99	3.96	2.72	3.94	4.92	3.51	4.16	3.62	3.37	12.27	11.79	10.04	9.59	40.48	From 1867 to 1874. Summary from eight years.
Ashland, N.H.	2.92	3.34	2.98	3.49	3.46	2.72	3.66	4.17	4.17	4.26	3.94	3.62	9.84	10.57	11.31	9.27	46.50	From January 1870 to 1874.
Bristol, "	2.86	2.70	2.47	3.18	3.42	3.89	3.41	3.79	3.17	4.01	4.57	3.87	9.57	11.09	10.56	9.08	46.08	From four years observation.
Claremont, "	3.41	3.55	3.10	3.69	2.48	2.88	4.66	3.38	3.61	4.38	3.94	3.50	9.52	10.83	13.07	9.52	44.08	From 1867 to 1866.
Concord, "	3.69	3.30	3.08	4.80	5.53	2.49	2.49	4.19	2.07	5.36	4.55	2.90	9.32	9.29	11.58	9.76	40.99	From 1849 to 1866.
Hanover, "	2.74	3.94	3.47	4.78	4.58	2.88	2.07	4.89	2.07	3.68	5.01	2.80	12.50	9.13	11.58	9.08	40.33	Reduced from 19 years observations at the College.
Lake Village, "	3.38	3.08	2.98	4.75	3.66	1.94	8.41	6.33	4.41	5.96	4.06	3.98	11.68	11.11	14.10	11.48	42.78	From January, 1870 to 1874.
Londonderry, "	3.69	2.47	2.47	.07	4.62	2.03	3.59	8.06	9.66	5.12	4.56	3.98	11.49	9.11	19.06	10.78	45.61	From 1849 to 1868.
Manchester, "	2.74	.29	3.94	1.01	3.02	18.46	4.32	6.33	3.98	6.10	3.02	3.96	5.52	28.37	10.63	9.33	45.17	From 1862 to 1864.
Meredith, "	1.67	2.47	4.01	2.48	3.31	2.73	4.92	3.41	5.38	3.76	2.76	1.88	6.07	8.88	8.04	2.81	44.08	From 1868 to 1875, inclusive.
Mt. Washington, "	3.39	2.53	3.82	3.02	2.45	3.94	4.69	3.84	3.98	8.64	3.21	3.08	9.26	11.50	10.83	7.44	55.79	From U. S Signal Service, 1871 and 1872.
Portsmouth, "	2.77	3.67	3.47	3.82	2.18	4.69	4.85	3.58	2.45	4.24	4.06	3.08	9.31	11.11	12.15	8.37	29.58	Four years observation on sea coast.
Stratford, "	3.63	3.63	3.84	3.82	2.45	2.87	4.85	3.84	2.45	4.90	3.66	2.44	9.57	11.96	10.31	10.31	38.96	From 1856 to 1866.
Wiers, "	3.53	2.31	4.01	2.88	2.18	2.87	5.57	3.84	5.64	4.51	3.95	2.05	9.41	12.66	12.10	8.78	42.86	From 1868 to 1873.
Wolfeborough, "	2.71	1.78	4.25	4.21	4.21	4.25	5.57	3.96	5.64	4.51	3.95	2.05	9.41	12.66	12.10	7.54	42.28	Same time as Wiers.
Fort Miller, N.Y.																	43.89	Four years,
Lansingburg, "	1.36	2.30	3.49	4.90	3.30	5.57	4.04	3.33	2.90	2.80	2.38	1.90	10.99	13.04	8.08	5.35	43.89	Monthly means and length of time not sent me.
Plattsburg, "																	37.41	Three years combined for result.
Troy, "																	49.07	Record lost, so no monthly means.
Amherst, Mass.	3.69	2.21	2.98	3.97	4.13	4.97	4.43	5.96	3.12	4.21	2.83	4.59	10.38	14.61	10.17	10.69	45.58	From 1868 to 1876.
Florida, "																	44.08	Monthly means not received.

Diagram 1 shows the fluctuations in the annual rainfall in the Atlantic States,—Maine to Maryland,—from 1805 to 1867. It will be noticed that there are groups of years of unusual amount of rain, followed by years of drouth; showing that one follows the other with something like regularity. As a whole it seems the amount of rainfall is increasing. The figures at the left are the percentage of the mean amount.

Diagram 2 shows the fluctuations in the annual rainfall at Lunenburgh, and also shows the same groups of years. It is never certain that a large amount of rainfall precludes the possibility of droughts, as it may not be well distributed through the year. The year 1871 was remarkable for its dry summer, yet the amount of rainfall is above the average.

Diagram 3 shows the fluctuations in the annual snowfall at same place. The difference is much greater than the difference in rainfall. The greatest amount, 167.5 inches, is more than twice as much as the mean, which is 83.1 inches, and the least amount is less than half the mean, being only forty-one inches. Still the grouping of winters is observable as in rainfall, yet shows no time where there was over three consecutive years above the mean amount.

Diagram 4 shows the annual fluctuations of rainfall at Lake Village, N. H., from 1857 to 1873. It will be noticed that while the grouping is seen, that it is not as marked, or the fluctuation as great, as in Connecticut river valley at Lunenburgh.

TEMPERATURE.

As the sun's rays are perpendicular at the equator, and strike at a regular increasing angle as we approach the poles, it would seem evident that there would be an equally decreasing degree of temperature, yet such is not the case. We find that latitude is by no means a sure criterion of heat and cold. The equator should thus be the warmest section of earth, but such is not the case, the warmest place on either continent being about ten degrees north of the real equator. Also the cold at the poles, not as might be expected at the poles themselves, but about fifteen degrees from them, and beyond that impenetrable ice barrier is open sea. Again the lines of temperature are by no means regular. England, six or eight degrees farther north than we are, has a climate warmer than New York city. And Quebec, on the St. Lawrence, has as cold a winter as the interior of Siberia; while Canada West, in higher latitude, is as warm as Massachusetts, and the winter is full six weeks shorter than at Quebec. Still the summer at Quebec is as warm as the summer of Italy, where snow never falls. Thus we see how the lay of the land, and the currents of air and of water effect climate quite as much as latitude. In a hilly State like Vermont the streams of cold in the winter are nearly as regular as the streams of water in the summer. The air in contact with the earth becomes colder, and for that reason heavier, and so flows down our valleys like streams of water. In the Connecticut river valley runs down a great stream of cold air, making quite regular temperature in winter the whole length of the State, and bending the annual temperature lines at least fifty miles further south in that valley than on the hills. I give a table extending through three years, from observations at Woodstock and Lunenburgh. These tables show that

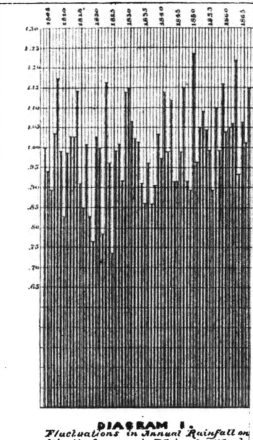

DIAGRAM I.
Fluctuations in Annual Rainfall on
Atlantic Sea-coast, Maine to Maryla[...]
from Smithsonian Rain Table, by C. A. Sc[...]

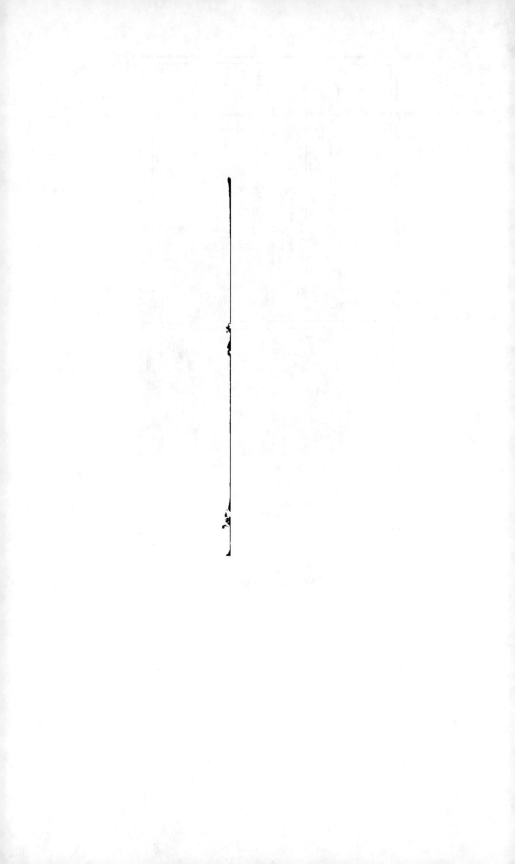

there is no great variation between them. Woodstock is 650 feet above tide water, and the place of observation in Lunenburgh is 1,210 feet above, being six hundred feet higher than the Connecticut river valley. The table also gives comparative rainfall. It is several degrees colder in winter in the river valley, two miles from my station, than here.

TABLES OF MONTHLY SNOW AND RAINFALL, MONTHLY MEAN, MAXIMUM, AND MINIMUM TEMPERATURES.

COMPILED FROM THE SMITHSONIAN AND OBSERVATIONS BY MYSELF.

1868.

		Jan.	Feb.	March.	April.	May.	June.	July.	Aug.	Sept.	Oct.	Nov.	Dec.
WOODSTOCK, VT.	Snow,—inches	.35	4.	.17	22.	1.	......	......	......	...,...	4.	16.	25.
	Rain, or melt. snow	3.50	.04	......	......	......	......	......	......	......	......	......	......
	Date	3	21	27	16	29	18	13	3,19 26	13	8	1	21
	Thermom., max	49	45	53	70	77	87.5	95	84	80	73	56	38.5
	Date	13	8	1	10	4	8	27	17	18	24	24	25
	Thermom., min	−20	−33	−28	−2.5	28	35	46	39	27	10.5	12	−27
	Mean	13.58	10.29	17.74	36.79	52.41	63.11	71.81	64.17	53.76	39.49	29.86	14.47
LUNENBURG, VT.	Snow,—inches	18.75	14.50	13.	5.50	......	......	......	......	......	......	......	......
	Rain, or melt. snow	1.87	1.45	1.50	1.30	4.50	4.40	4.62	1.81	8.08	1.30	7.15	2.45
	Date	2	20	31	16	28	18	13, 15	3	12	7	1	21
	Thermom., max	33	37	57	68	77	87	75	85	80	65	54	32
	Date	13	24	1	10	9	3	28	17, 28	18	30	17, 30	27
	Thermom., min	−16	−27	−12	4	32	36	53	49	25	15	14	−30
	Mean	12.2	8.78	27.93	34.	52.48	62.15	71.58	66.73	54.58	41.6	29.60	13.82

1869.

		Jan.	Feb.	March.	April.	May.	June.	July.	Aug.	Sept.	Oct.	Nov.	Dec.
WOODSTOCK, VT.	Snow,—inches	22.	40.	26.	2.	5.	......	......	......	......	1.50	6.	14.30
	Rain, or melt. snow	......	......	......	......	......	2.85	1.76	2.24	3.44	12.90	2.29	2.94
	Date	8	13	27	27	12	4	25	20	7,8,20	14	4	1
	Thermom., max	50	50.5	53	66	85	84	86	88	82	73	59	53
	Date	23	8	1	11	1	7	6	8	28	28	26	8
	Thermom., min	−25.5	−21	−28	17	23	32	40	35	31	18	7	−18
	Mean	19.06	19.37	17.74	39.71	52.51	59.84	65.22	61.59	59.12	41.96	30.	22.44
LUNENBURG, VT.	Snow,—inches	18.	36.	19.5	3.	1.	......	......	......	......	3.5	2.50	7.75
	Rain, or melt. snow	1.95	3.67	2.5	1.75	2.75	5.	2.75	2.50	2.60	7.72	2.	2.25
	Date	13, 19	13	28	20	31	4	11	25	17	1	5	1
	Thermom., max	35	39	47	61	78	84	84	78	82	72	56	40
	Date	8	8	2	5	1, 4	6	1	31	28	28	26	8
	Thermom., min	−20	−12	−23	20	34	46	52	48	36	21	13	−12
	Mean	14.73	19.80	19.43	38.25	53.88	60.93	67.58	63.43	61.60	42.90	30.53	21.93

1870.

		Jan.	Feb.	March.	April.	May.	June.	July.	Aug.	Sept.	Oct.	Nov.	Dec.
WOODSTOCK, VT.	Snow,—inches	25.39	36.50	24.50	3.50	......	......	......	......	......	4.	2.	26.75
	Rain, or melt. snow	6.90	5.12	2.60	3.27	1.81	5.35	1.82	1.08	4.98	3.80	1.87	2.51
	Date	28	12, 18	30	28	16	25	24	9	1, 2	16	9	28
	Thermom., max	52	45	57	78	83.5	91.5	90	91.5	85	70	61	51.5
	Date	14	5	4	1	6	23	2	27	18	27	16	21
	Thermom., min	−5	−14	−8	18	29.5	49	43	34	31.5	17	15	27
	Mean	23.60	16.07	23.66	42.04	54.05	68.04	70.10	65.61	57.01	45.71	34.17	16.49
LUNENBURG, VT.	Snow,—inches	23.05	40.	17.25	......	......	......	......	......	......	.50	9.	8.
	Rain, or melt. snow	4.55	4.10	4.52	2.50	4.00	3.5	4.55	6.42	3.	3.95	5.25	1.30
	Date	17	18	31	27	30, 31	29	24	9	4	12	9	2
	Thermom., max	42	44	52	68	82	94	89	90	80	79	56	40
	Date	14	2	12	4	7	22	2	26	12	27	27	25
	Thermom., min	−12	−18	−15	25	35	50	56	48	38	17	18	−18
	Mean	21.88	16.28	23.73	42.70	53.43	68.68	72.57	67.25	59.75	46.20	33.55	21.55

For this period the heat of summer has not exceeded 100 degrees, and but twice has the thermometer touched that point. It has once touched 45 degrees below zero. The extreme heat of summer rarely exceeds 90 degrees, and the cold of winter is seldom greater than 25 degrees below zero.

The average number of stormy days in a year for this period has been 120, and the average number of fair days 118, of cloudy days 127. From this it is reasonable to infer that our time is pretty equally divided into fair, cloudy, and stormy weather. In summer we have an excess of fair weather while in winter there is an excess of stormy or cloudy weather.

THE ATMOSPHERE.

It is desirable that all should have a general understanding of our atmosphere, and the laws by which our storms are regulated or produced, and to render such instructive I shall say something of the history of Meteorological Science, and also of familiar signs as well as instrumental observations. Two hundred and fifty years ago it was not known that we had an atmosphere. All the phenomena it produces were explained upon other principles, some of them showing the wildest theories and the most absurd ideas. The creation of the atmosphere as declared in Genesis, as the "firmament" dividing the waters, was not understood. A vague and unmeaning explanation was given it. When it was discovered that there was in reality an æriform fluid surrounding the earth, possessing weight, color, power of diffusing light and heat, and necessary to the existence of all animal and vegetable life, it struck with wonder and astonishment all the learned throughout the world. So wonderful and incredible did it at first appear, that it was not until after the lapse of several years, till opinions which had prevailed for ages were overthrown, and the most decisive experiments had been performed in every possible way, that it was cordially received. This atmosphere is composed mainly of two gases, nitrogen and oxygen. It was, however, less than eighty years ago, supposed to be a simple body, but is now known to be composed of about eighty parts by measure of nitrogen, and twenty parts of oxygen. It may be necessary to offer a few remarks on these gases, opposite in their nature ; entering into no chemical union, yet being combined in so exact proportion as to support animal and vegetable life, and the smallest change, perhaps, detrimental to either. Oxygen gas is eminently the supporter of combustion, and ignited substances burn in it with the most intense brilliancy. Even shavings of zinc and iron may be ignited, by dipping the ends in melted brimstone, and introducing them into this gas while the brimstone is on fire. They then burn with intense heat and give a peculiar light, exemplifying the fact that if our globe was surrounded by an increased amount of oxygen, many now incombustible substances could be burned.

Nitrogen gas is exactly opposite in quality. It will extinguish fire as well as water, and will soon kill any animal that breathes it uncombined with oxygen. Yet four fifths of the air we breathe is this noxious substance.

Oxygen is the life-giving element, and as this is largely consumed in combustion and respiration, and by those processes replaced by an equal volume of carbonic acid, which is detrimental to animal life, it would seem that the atmosphere would at length become deleterious. This would be the case, were it not for vegetation, which by aid of the sun's rays, absorbs the carbonic acid, and gives off, after the appropriation of the carbon, oxygen for the animal. Thus the animal and vegetable mutually support each other; I say support, as breathing affords three fourths of our own nourishment; leaving the other quarter, only, to be supplied by food. With this unceasing metamorphosis in beings and things, goes on a continuous exchange, by virtue of which the gases of the atmosphere take up their abode in animal and plant. Each atom of air, therefore, passes from life to life as it escapes from death after death, being in turn wind, flood, animal, plant, or flower, being successively employed in the composition of thousands of plants and animals.

It is the inexhaustible source from whence everything that lives draws much the largest share of its support, and into which everything that dies contributes. Under its action vegetables and animals are brought into existence and then perish.

Life and death are alike taken in at every respiration, and the atom of oxygen which escapes from the blade of grass may find its way into the lungs of the infant in the cradle; or the last sigh of a dying man go to nourish the brilliant petal of a flower.

WIND AND STORM.

We know wind is air in motion; this of course could not be known when it was not known that we had an atmosphere, so by the ancients entirely different causes were supposed to produce it.

Pliny, one of our greatest historians, being thoroughly reliable as regards the facts of his day, says, "In houses there be hollow places devised and made by men's hands for receipt of wind, which being enclosed with shade and darkness, gather their blasts." Thus in the time of the ancient Greeks; in the time of renowned men that invented an architecture, or manner of finish, which has been handed down and used to the present day, building their houses and their stately edifices, with curiously contrived cavities on the outside, as a resting place for wind. Wind to their minds was a spirit needing their protection, and one to which they did not wish to give offense. They supposed his abode to be in the mountains of the North which divided their country round about the Mediterranean, from the country of the fabled Hyperboreans, in the North, and that when displeased they hurled down the North wind, which was chilly and cold to the inhabitants of a more Southern clime.

Caves were places of superstitious dread, as they were supposed to be the resting places of wind, and finally became of use to the freebooters, as they were thus comparatively safe from the law. We learn from many ancient histories similar facts. The origin of wind as taught to their youth, and handed down to us in their ancient manuscript school books is this: "There be certain caves and holes in the earth which breed wind continually without end. They have wide mouths, and if you cast in anything of light weight, it will be seen presently to come out with a stormy tempest. Thus you can see how all winds have a cause." That all winds have a cause we are ready to admit, but that they understood that cause no one can now believe. Even today we are in doubt about some of the causes of atmospheric disturbances. We know, however, that heat and electricity, as well as the revolution of the earth, and atmospheric tides, aid in the development of winds and storms.

Though all parts of the earth's surface have atmospheric currents of more or less force, there is great contrast in different sections. The Northern half of the Connecticut River valley, shut out by the White Mountains on the East from the oceanic currents, and by the Green Mountains on the West from the strong West winds of the continent, have aerial currents strictly local, and of mild force. The winds prevalent around Lake Champlain even, would be considered extraordinary gales here, while in many sections of our fertile West, the wind blows every day so as seriously to impede labor, while in many places peculiar winds and violent hurricanes pass, frequently causing great inconvenience, and many times fearful loss of life and property. Though our quiet State is seldom visited by tornadoes, there has been a sufficient number to show that we are not exempt. Upon the 3d of July, 1842, there was a tornado in the town of Victory of remarkable force. Its devastations commenced upon the top of a high hill in that town, where its path was only a few rods wide, but it gradually increased to about half a mile in width, sweeping everything before it for about two miles, when it seemed to lose its power. Its track was a forest, yet it not only tore up the trees, but the soil also, piling it with the twisted and broken trees in huge rows near the place where its fury seemed spent. The noise of this tornado was heard for more than ten miles, and was supposed by many to be an earthquake. It was accompanied by heavy thunder and incessant lightning, with torrents of rain.

While tornados are rare, hail storms in some special sections of the State are of more frequent occurrence, and are usually electrical. But what electricity has to do with the formation of hail storms is not so clear. All we know of the nature of electricity is that it is a mighty force, called for the sake of convenience, a fluid. It appears to exist in every substance in nature, in both solids and fluids, and may be roused from its repose by a variety of causes, such as friction, heat, and chemical action. But we are totally ignorant of the reason why these causes excite it. When thus aroused it often displays tremendous power, and sometimes produces the most destructive effects.

In the formation of a thunder shower two or more clouds will first be seen, and seeming to attract each other, they approach within a certain distance, when their electricity begins to accumulate on the sides nearest each other. When this accumulation becomes sufficiently intense to overcome the resistance of the atmosphere, discharges take place, causing thunder and lightning. The same also frequently takes place between the clouds and the earth, in which case the lightning may pass both upward and downward.

The lightning is first seen, because light travels with immense velocity, while sound travels only 1142 feet per second. Thus it may be seen that if four and one half seconds intervene between the flash and the report, one mile intervenes between the discharge and the observer. In this way the distance can always be ascertained unless the cloud is more than ten miles distant, in which case, in ordinary circumstances, no thunder will be heard. This sublime and terrific phenomena is well known to every individual, as it is occasionally displayed in every region of the globe. A thunder storm usually happens in calm weather, though they are sometimes accompanied with furious winds. The dust storms of Africa and India are also accompanied with much thunder and lightning. There is a peculiar circumstance sometimes attending thunder showers that converts them into hail-storms as before mentioned. The clouds in some cases seem to be very high, or carried up by the wind beneath, until, from their elevation, the water they contain may become frozen, and fall to the earth as hail. They are sometimes frozen very rapidly, the hail falling in all possible shapes, but generally the hail-stones are round or oblong in shape, and when broken open seem to be frozen round a snowy nucleus, frequently as well formed as a snow-flake, being star-shaped, like a large variety of our snow storms, showing that snow was first formed from a part of the water, every flake forming a nucleus for the formation of the water around it into a hail-stone. In other cases the hail seems to result from a purely electrical condition of the clouds, which in some way furnishes the cold necessary to freeze the rain drops.

The atmosphere aided by various disturbing powers resulting in wind, and its remarkable avidity for water known as evaporation; distributes moisture to the earth as dew or rain, rendering it fertile wherever this moisture is deposited. The most simple form of deposition is that of dew. This is the humidity of the atmosphere deposited on surfaces with which it comes in contact. During the day the solid portion of the earth becomes more heated than the atmosphere, and gives off moisture by evaporation. This continues to go on until the solid portion becomes cooler than the air, then this moisture is in part condensed on the solids, forming dew. Dew can be artificially formed at any hour of the day, as for instance: Take a pitcher of ice water, or cold spring water, and place it in the open air, or in a room in your dwelling, and the outside of the pitcher will in a short time become covered with drops of water. The general term given is that the pitcher sweats, and some even suppose that the water sweats through the pores of the pitcher. But such is not the case; it is dew, the same as deposited on the grass or ground, and from the same cause. Warm days and cool nights are favorable for large dews. I might almost say fair nights, for cloudy nights are always cooler, and sometimes, and in fact often, cool the atmosphere as fast as the earth cools, in which case there can be no dew. This shows an atmospheric change likely to be followed by rain. The atmosphere always contains more or less aqueous vapor in an invisible form. This vapor is water dissolved in it, in the same manner as refined sugar can be dissolved in water without changing its transparency. As warm water will dissolve more sugar than cold, so will a warm atmosphere dissolve more water, or in other words retain more vapor. If the temperature is depressed, the vapor appears in the form of clouds. If greatly cooled it will fall as rain, hail, or snow. This principle of evaporation and deposit of moisture now so generally understood and so thoroughly demonstrated, was once like all other points in meteorology considered as beyond comprehension; and attributed to supernatural causes. Horace speaks of dew as a gentle evening shower without clouds. Virgil says that every night we have a misty rain. Pliny, of the falling dew; and even at the present day how often do we hear the expressions about falling dew, showing that though we understand the principle and manner of deposit, we are not entirely free from the idiom of expression founded on the mythic reasons of past ages. Some countries are rendered productive and even fertile by dew alone, and doubtless this to the ancients being so mysterious, being deposited in greatest abundance when most wanted, and of vastly more importance to them than to us, caused them to regard it as an express gift of God, as the manna in the wilderness, by his hand, but not in accordance with the laws of nature. In the minds of those people it possessed wonderful virtues. It was supposed if man sipped the dew, and drank no other beverage he would possess new vital energies, and live on the earth a very long period, if not forever. Even up to the present time it is considered by some to beautify the complexion and restore the charms of youth. While by the alchemist of olden time it was supposed to be a solvent of mysterious power. In fine almost everything possible or impossible has been ascribed to its many virtues. In ancient philosophy it is described as the tears of the gods, and by all ancient nations as a gift from heaven. Thus we can see why, when the blessing of Isaac was conferred upon his son Jacob he used the expression, "God give thee of the dew of heaven." In the East it was much more abundant and important than with us, although of great importance here.

It begins usually to be deposited as soon as the sun goes down, and frequently before, and continues through the night. When the temperature decreases during the night to the freezing point, the dew is frozen, forming frost, its destructive powers being in proportion to the intensity of the cold. The glistening appearance on the surface of snow, are dewy particles frozen into beautiful prisms; when the sun shines upon them in the right direction they will reflect all the colors of the rainbow. Dew is deposited in all the fertile regions of the earth, but in sandy deserts, where the heat of the sun is so intense, the burning sands are never cool enough for its deposit. A desert is, therefore, not only a rainless tract, but a dewless one. This deposit not only varies greatly with the season, but in different sections much in amount.

The dew in this section amounts to about three and a half inches of water per annum. In Europe to about five.

Dew is, as we have seen, the most simple form of disseminated moisture, yet simple as it is, it is very efficient. Indeed, every shrub and herb, every leaf and blade of grass, possesses according to its wants a different power of radiation, so that each condenses as much dew as is necessary for its own individual wants. Thus not even a single dew drop seems to have been formed by the rude hand of chance, but it is adjusted by the balance of infinite wisdom to accomplish a definite and benevolent end.

CLOUDS.

The next means of disseminating moisture is by clouds. Their formation I have already explained. They are not vapor, as vapor is invisible, but water. Not held in solution, but in minute particles, supposed to be hollow globes, as fine as floating dust. In reality, water pulverized,—so light that they may be readily blown forward by winds, yet ready to be condensed and precipitated as rain by the slightest causes. This condensation is commonly produced by reduction of temperature, either by a cooler current of air or by electrical changes. Clouds are continually varying in their form and appearance, but may be classed under four heads.

THE CIRRUS

is a light fleecy cloud resembling a wisp of hair or bunch of feathers, and is the highest of all clouds. The water it contains is of course frozen into snowy particles on account of its great elevation; and further, we know it is so, as æronauts passing through it in their balloons always find it so, and its peculiar appearance is owing to that fact, as the wind blows the frozen particles about in long wavy lines. When this cloud is seen, if it be watched, it will gradually change into a sort of dappled sky or wavy cloud, and disappear, or else the length of the lines will increase and expand over the heavens; in this section the resolution into the mottled sky indicates fair weather, but the last mentioned form indicates a storm, which will be likely to come in three days or less. Still when we have had a storm, and those clouds are seen, it is a sure indication that the storm is about to abate, and for that reason they are termed by some fair weather clouds. Second,

THE CUMULUS

is the common cloud wafted about by the wind, sometimes piled up in the horizon so as to look like snow-capped mountains, or like ocean billows lashed with foam. If those clouds move off and appear high above the mountains, fair weather may be expected, but if the lower edges are smooth, and they hang around our mountains, look out for storms, for they then have to settle but a few feet nearer the earth to become rain clouds. In summer they produce many showers, especially when the weather is hot; as evaporation is rapid they increase in size, and sinking nearer the earth they discharge their contents with thunder and lightning. You can usually tell something of the state of the atmosphere and probabilities of a storm by watching a small cloud of this description. If it increases in size and assumes a darker tint, it indicates a storm; if, on the other hand, it decreases, it indicates fair weather, and if it be entirely dissipated no storm may be expected, but if dissipated rapidly look out for showers, as the ' electrical condition of the atmosphere is such that they may be produced as soon as evaporation supplies the requisite amount of moisture. Third,

THE STRATUS

is a horizontal misty cloud, frequently observed in a summer evening lying across the country, slightly elevated, and apparently having no motion, also along rivers and over ponds where it is known as fog. It is usually dissipated by the sun in the early morning, but sometimes is in sufficient quantity to produce rain, but in this section is generally a fair weather cloud. Fourth,

THE NIMBUS

or rain cloud has a uniform gray color, has fringed edges when they are seen, but usually covers the entire sky, and the rapidity of the rain depends on its thickness. The nearer the ground the greater its probable thickness, and the more water contained within it. All know this cloud, in a moment, as it always produces rain. With us it is a common cloud, but in many sections of our globe it is never seen, as extensive tracts of land exist where rain never falls. Some of this land is rendered productive by the overflow of rivers like Egypt, others by copious dews like some parts of Arabia, but by far the larger portion of the rainless districts are barren deserts like Chili and Peru, or the great desert of Africa or Arabia. On the other hand within the tropics where the great cloud belt that continually encircles the earth swings gradually North and South, changing their seasons from wet to dry, there are places where rain and mist continually fall, but as those tracts are among islands or upon the ocean, they are of little account. The island of Tahiti forms a striking example, rising out of the tropical ocean to ten thousand feet above its surface, it forms a nucleus for the condensation of clouds, and the thickly wooded sides are well fitted for rapid evaporation by the hot winds of the ocean, so there goes on with great power rapid evaporation and incessant rain. The amount of rain which usually falls in Vermont, in a rainy day, is seldom over one inch, and generally much less, the entire fall of water including melted snow, is on an average about forty inches. We are, however, like other countries, liable to very great

IRREGULARITY OF STORMS,

and on some rare occasions as many as four inches have fallen in twenty-four hours, as on the 3d of October, 1869. With this exception we have not for twenty years had over two inches in twenty-four hours, and that only seven times during this period. These were all remarkable rains. As it takes a foot of snow to make an inch of water, you can see our winter storms are much less than those of summer. Of remarkable storms I will notice, at Catskill, N. Y., July 26th, 1819, eighteen inches of water fell in seven and a half hours, making a flood which swept everything before it. Great numbers of cattle and many buildings shared the devastation. Whole forests were uprooted and floated away. In California, July, 1862, thirty-two inches fell in twelve hours. Though this was only the case in a small section of country, the ruin was almost without precedent. Large brick buildings were carried away so the least trace of them could not be found, old beds of rivers were filled up and new formed, and devastation was the order of the day. This, in fine, was only one of three floods that has visited that country since it became a part of the United States. At Genoa, Oct. 25th, 1865, thirty-four inches of rain fell in twelve hours, causing general destruction of property. These floods in temperate zones are rare visitations, but not so in the torrid, as hardly a single day passes but what a tornado, or deluge, takes place on some limited portion of that section. Sometimes as many as thirty-five inches of water fall in eight or nine hours, as was the case in the Cuban tornado of 1867, which swept away almost everything on the island, reducing the inhabitants to a state of starvation, donations being sent out from New York and other cities for the thousands of sufferers. Many instances are recorded of fish, frogs, and a great variety of animal and vegetable substances, falling from the clouds.

These are sometimes taken up by waterspouts and whirlwinds, and in countries where those are prevalent are of no rare occurrence. Again, substances resembling flesh form in the atmosphere, as in the Kentucky meat shower of 1876, which was this peculiar substance. Or like the vegetable manna that fell in Russia in 1817, in sufficient quantities to afford food for cattle, and even for man, and form an article of traffic. Between twenty and thirty showers of blood are recorded in history, but when this blood has been examined by scientists, it has been found to be red earth mixed in the rain drops, and of course cannot be unusual where the bare red earth covered with soil impregnated with iron, is exposed to violent winds, capable of raising the fine particles to be brought to earth in raindrops.

Showers of sulphur, so often recorded, are almost always the yellow pollen of the birch or pine. But we will leave these curious freaks of nature, and look more particularly to our own climate.

The Smithsonian Institute, with a corps of observers throughout the world, endeavored to establish a theory by which coming storms could be predicted. Though they failed in accomplishing all they hoped to do, yet it became evident that by a series of observations taken in different sections, by competent observers, much might be gained, and by telegraphing such observations to a central point the progress of storms could be known; and the telegraph could convey in advance the probable force and effect of coming storms. It was at once seen that such an arrangement would greatly benefit the shipping off our coasts, and the agriculturists of our country; and the War department in connection with the Postoffice department, organized a service for the purpose of conveying such information over the country for the benefit of all. Though the appropriations have. been entirely inadequate for the work, much has been accomplished and millions of dollars saved.

As I have before stated, most of our storms commence far south or west. We frequently receive notice from the Signal Service, stating that such a storm is coming, when the heavens are unclouded, and nothing but a slight decline in the barometer indicates its approach. Many of those storms expend their fury before reaching us, and a cloudy day only will show their previous existence.

HOW TO FORETELL STORMS.

To the agriculturalist the ability to foretell, even for a day, the approach of a storm, is many times of great advantage. I much regret that all are not able to derive that knowledge from the Signal Service reports, but as such cannot be at present, I will explain the uses of

3

such instruments as will aid in forming a reliable conclusion as to the probabilities, many times amounting to an absolute certainty, of a coming storm, and then mention the reliable signs, without instruments, which my observations have established as worthy of note.

First, it is necessary to ascertain the amount of water held either by saturation or clouds in the atmosphere. If the atmosphere in this latitude is thoroughly saturated, it may contain about eight inches of water, all invisible to the general observer, yet with such a degree of saturation the heavenly bodies will appear somewhat dimly, and as water is a much better conductor of sounds than air, sounds will be more readily transmitted. People will notice the difference, and call it the sign of a storm. It shows a fact, that is, that a large amount of moisture is contained in the atmosphere. But as without instruments the amount of moisture must become a Yankee guess, it will not answer for the scientific observer. He must have something more tangible.

With the common uses of the thermometer all are acquainted. At first thought it would seem impossible that the thermometer could be used to show the exact amount of moisture, as well as heat, in our atmosphere, yet such is the case. If we hang two thermometers side by side, they will indicate the same temperature; but if we cover the bulb of one with fibres of silk, and saturate it with water, we shall find that the evaporation of the water will produce a greater degree of cold, and that thermometer will fall, sometimes as much as 8° or 10°. Now it is found that when the thermometers stand several degrees apart that the atmosphere is very dry, and consequently evaporation rapid; but as it becomes saturated, and evaporation less rapid, the thermometers will more nearly coincide, and when evaporation can go on no longer, the atmosphere containing all the moisture it can, they will then be exactly alike; but of course rain must be imminent. This instrument when put up in proper form, with tube to contain water to keep the bulb of one thermometer constantly wet, is called a psychrometer, and is an important instrument to the meteorological observer.

Tables have been constructed so that the exact amount of moisture contained in the atmosphere can, in connection with this instrument, be always known; but as a large amount of moisture may be long retained unless there is some decided change, we have to look to the barometer for guidance. With this instrument there is one general direction, which if allowed to have full scope would obviate many difficulties. That is, the rise of the mercury from any point denotes less wind, and a general improvement in the atmosphere, with less inclination to storm. A fall from any point indicates wind and storm, or a condition of the atmosphere more favorable to such. But as wind, rain or snow affect the barometer in the same manner, how are we to prognosticate which will take place? Here we receive aid from the psychrometer. If, when the fall takes place, the atmosphere is fully saturated with moisture, of a necessity this moisture will, in part at least, be condensed and fall; if, on the other hand, the atmosphere contains but little moisture, wind only will result, yet the fall may be equally great. By this you will see that those two instruments go hand in hand if you would form a good judgment. Yet to the per-

son possessing only a· barometer, I would say that a due reference to
the direction of the wind, the formation of the clouds, and the temperature of the atmosphere, will give sufficient data upon which to
predicate a reliable opinion.

SIGNS OF RAIN; VISIBLE AND INVISIBLE.

A hunter ere the sun dispels the fog of a mucky morn, builds his
camp-fire in the forest. The stifling smoke arises from damp and half
decayed fagots. His culinary duties release the fragrant odors, from
eggs and bacon, while frying onions still send forth their aroma,
mingled with the sweet williams, violets and blue bells, at his feet.
His dogs sniff the air, and from this medley of smells scent a fox, and
start upon its trail, while another follows a rabbit. In this is a lesson
for the thoughtful, humidity, smoke, stench and odors of a hundred
kinds, arise and impregnate the atmosphere which previously occupied
the whole space; and many contend that the relative gases in the atmosphere are unchanged. Then we have the anomaly of a vessel a
hundred times full without once emptying. This also illustrates how
the atmosphere may become, all invisible to us, the vehicle for the con-
veyance of moisture.

It has been ascertained by experiment that when the atmosphere is
fully saturated, it may contain as much as thirteen grains of water to
the cubic foot; all invisible to the naked eye, if at eighty-six degrees
of temperature. Now, how shall we ascertain its presence? The
barometer and psychrometer aid us but are not always at hand. Common observations have established in the minds of all certain signs of
storm, which have in many cases a foundation in scientific fact. The
following lines first published many years ago, and emanating from an
unknown pen, embody many of those, and so I reproduce it as a text
upon which I can offer explanations:

        " The hollow winds begin to blow,
        And the barometer is low.
        The soot falls down, the spaniels sleep,
        And spiders from their cobwebs peep.
        Last night the sun went pale to bed,
        The moon in halo hid her head.
        The clouds look black appearing nigh,
        And see, a rainbow spans the sky.
        The walls are damp, the ditches smell,
        Closed is the pink eyed pimpernel.
        Hark how the chairs and tables crack,
        Rheumatic joints are on the rack.
        Loud quack the ducks, the sea fowl cry,
        The distant hills are looking nigh.
        How restless are the snorting swine,
        The busy flies disturb the kine.
        Low o'er the grass the swallow wings,
        The cricket, too, how sharp he sings.
        Puss on the hearth with velvet paws,
        Sits wiping o'er her whiskered jaws.
        The smoke from chimneys right ascends,
        Then spreading back to earth it bends.
        The wind unsteady, veers around,
        Or, settling in the south is found,
        Through the clear stream the fishes rise,
        And nimbly catch the incautious flies.
        The glow worms numerous, clear and bright,
        Illumed the dewy hill last night.
        At dusk the squalid toad was seen

Like quadruped, stalk on the green.
The whirling wind, the dust obeys
And in the rapid eddy plays.
The frog has changed his yellow vest,
And in a russet coat is dressed.
The sky is green, the air is still,
The mellow blackbird's voice is shrill.
The dog so altered in his taste,
Quits usual food on grass to feast.
Behold the crows, how odd their flight
They imitate the gliding kite,
And seem precipitate to fall,
As if they felt the piercing ball,
The tender colts on back do lie,
Nor heed the traveler passing by.
The rocks are seen in shade to sweat,
Although the sun is radiant yet.
In fiery red the sun doth rise.
Then wades through clouds to mount the skies.
'Twill surely rain, we see it with sorrow
'Twill rain today and rain tomorrow."

As the "hollow winds" are South winds, which are productive of moisture, and are likely to cause a depression in the atmospheric ocean, the barometer usually falls when they blow, and air of a cooler temperature coming from the North to fill up that depression, usually condenses the moisture, causing rain. The "soot falls down," as by the absorption of water from the overcharged atmosphere, it is loosened and crumbles off.

Spiders on the alert for food, build their webs to catch flies; and as their hunger, like that of flies and fishes, becomes more intense when a part of the atmospheric pressure is removed, they are very impatient for food, and so peep out from their hiding places. Who has not noticed spider webs in the fields flat, and also at different angles, and heard the remark that when flat they denoted fair weather, and otherwise foul weather. The spider, intent only upon his food for the ensuing day, spins his web with instinct, almost reason. A fine day is always preceded after midnight with perfect calm, and at this time the spider builds his web flat. If a change of weather is in progress, this calm does not exist, and a breeze is in its place. The spider, knowing that rain-drops break his web, if flat, builds it at an angle according to the amount of wind, so the rain-drops will not strike fairly and break it. They sometimes fast, not building at all when the storm is to be very violent.

Circles around the moon, or mock suns, known as sun dogs, show that there is a large amount of moisture in the higher atmosphere, and are always brightest upon the sides of the sun from whence the wind is blowing in the upper atmosphere. As an illustration, we reason from this phenomenon thus: It is winter, the sun dog is on the south side of the sun, or brightest on that side. It shows south wind with moisture, which indicates warmer weather, with rain or snow. If upon the north side, cold winds with a small amount of snow.

Rainbows are formed by reflection of the sun's rays from falling drops of rain. If it rains in the morning when the sun's rays should dissipate the clouds, if a fair day, it is quite likely to rain more or less all day. From this the saying so common,—

"A rainbow in the morning,
Sailors take warning;
A rainbow at night,
Sailors take delight,"

is as a general thing correct, showing foul weather if in the morning, and a pleasant night, and perhaps for the next day, if seen at night.

"The walls are damp, the ditches smell."

In fair weather the atmosphere is usually more dense and the barometer higher, which shows the atmospheric pressure is greater ; the gases which escape from foul places, rise high above our organs of smell, as their specific gravity is nearly uniform, while before a storm the atmosphere is less dense ; consequently, as the disagreeable gases are the same heft, they float within range of our organs of smell, or but little, if any, higher than our heads. Thus they show by smell, probable storm.

" Closed is the pink-eyed pimpernell."

There are five plants which have been observed from time immemorial as indications of weather changes. They are the dandelion, trefoil, pimpernell, chickweed, and Siberian thistle.

The down of the dandelion closes for bad weather, but expands upon the return of sunshine. The trefoil is observed in Europe to have swelled stalks before rain, and contracts its leaves before showers. The pimpernel, which I suppose is the *anagallis arvensis* of Linnæus, is found in this country, and in some States grows in old fields, and among grain, where I have heard it called "devil weed." It flowers all summer, and when found in the morning with its red flowers shut, it will be a foul day ; when fully open, a fair day ; partially so, doubtful weather. Chickweed, which is common the world over, is an excellent weather guide. When its flowers expand freely, no rain may be expected for some hours; if it continues' open for twenty-four hours, no rain for days. Half open all the time, showery uncertain weather ; closed, long continued rain. If the flowers of the Siberian thistle are open all night, rain the next day is probable.

As regards the effect on animal life, what the coming storms have to do with it, is not so clear; yet the weather prognostics derived from the animal kingdom are often more to be relied upon than many people imagine. The lassitude and indefinable anxiety of nervous individuals before a thunder storm, the shooting pains in old wounds or scars, the aching and pricking of chilblains and corns, the attacks of headache or rheumatism, when a considerable change of the weather is about to take place, entitle many sensitive persons to be called living barometers, and so with the animal kingdom. I am inclined to think that the changes in atmospheric pressure may have to do with it, as it certainly does with the appetites of fishes and insects. It is well understood that fish generally bite best before a storm, and it is then that fishermen are about. They bite the best, as they feel a peculiar hunger when the barometer is low, showing less pressure from our atmosphere, and that is usually before a storm.

It is the same with insects. It would seem that at such times they had extraordinary feelings of hunger. All have realized, more or less, their greediness before a storm.

"Low o'er the grass the swallow wings."

It is a well known fact in natural history that certain insects have an exact number of vibrations of the wing to the minute, and if they fly at all it is in the same manner. When the barometer is high, the atmosphere near the surface of the earth is more dense, from the increased pressure, and such insects consequently fly as high or higher than our buildings. With less pressure, or a low barometer, they will rise but a few feet above the ground.

The swallows that feed upon these insects, which they catch upon the wing, fly at the height that their food flies, which, on account of their diminutive size do not attract our attention, while the swallows readily do. So even this sign has scientific fact for its basis.

Thus I might follow out line after line of the poem, showing its agreement with well known facts; but I should risk tiring your good nature in so doing, and as I have given you examples, will only add that the sweating of rocks, so called, being frequently supposed actual sweat, is only a deposit of dew; and when deposited from the atmosphere in the daytime, shows that there is a very large amount of water held in saturation, and a change is likely to occur, causing it to fall as rain.

Rain may be expected when the sun rises pale, or when, after setting, clouds ascend in the part of the sky from whence the wind blows. A red morning sky is usually followed by rain; a red evening sky, by fair weather. For twenty-five years the probabilities in these cases as true, have been as ten to one. This proves the old adage—

> " Evening red and morning gray,
> Will help the traveler on his way;
> But evening gray and morning red,
> Will bring down rain upon his head,"

was founded on fact. When the moon is pale in color, with horns blunt at first rising, the stars seeming large, dull and pale, with no perceptible twinkle, or if encompassed with a circle, it indicates an excess of moisture in the atmosphere, and probable rain. If mists are attracted by tops of hills or mountains after clear weather, expect rain in a day or two. In dry weather, if mists ascend from woodland or water, higher than usual in the morning, look out for immediate rain; but if such mists appear after sunset or before sunrise, it denotes fair and warmer weather the next day. If the rain commences an hour or two before sunrise, the clouds will probably be dissipated by the heat of the sun, and it will clear up before noon. If it commences an hour or two after sunrise, it will be likely to rain all day. A bright moon with sharp horns, or sharp definition if full, wind shifting North, a sky full of bright twinkling stars, small clouds at North and none at South, and perhaps appearing in heaps, denote colder weather. The same with clouds and winds shifting South, indicate warmer weather.

Whether a storm will clear off warm or cold is often a question of moment, and must be ascertained from experience and observation. From a long series of observations I would say that generally when the clouds break up so that we can catch glances of the cirrus cloud, if that is seen to move from the south or the southwest, the storm will clear off warm; if from the opposite direction, it will clear off cold. Again, if the surface winds wheel round from the south to the west, warm weather is indicated; if from the north to the west, it will be cooler. So in time of thunder showers, if they pass to the south of us, the atmosphere will seem much cooler than if they pass to the north. Again, the waves of temperature approach regularly. The old adage, that one extreme follows another, is verified in fact; and we usually have gradually increasing cold followed by as regular an increase of temperature. This is more readily discernible in the winter season, as will be seen by every one who has a thermometer, and will take the trouble to look at it. It is during this season that we have the greatest atmospheric changes, sometimes even as great as 100° of temperature in twenty-four hours.

There is another class of signs which some believe in, that are merely superstitions, having no foundation in fact. The Hindoos have their rain gods, the South Sea Islanders their wind conjurers, and the negroes of Africa their rain doctors; and previously we had our weather Almanacs and our Herschel's weather tables, one just as good as the other. Not many years ago that celebrated Herschel's weather table, which Herschel never saw, was considered almost infallible, and Thomas' Almanac quite so; but all enlightened people, unless some whose age has outgrown science, discard them. For the last thirty years our storms have taken place without regard to moon's quarters. We have had 2,668 storms, divided as follows: at new moon, 660; first quarter, 664; full moon, 668; last quarter, 676. This shows very plainly that the moon has nothing to do with storms. If the generally received idea was true, what little difference there is goes directly against it. The truth is that the moon has so little, if any, influence upon the weather, that men have never found out which way it is; and I can say, without fear of contradiction, that rain and wind doctors, and Hindoo gods, have just as much to do with the weather as this weather table, and no more.

In snow storms the form of the flakes, if observed, will tell the probable amount of the storm. Without cuts a description of the shape would be useless, and I therefore ask all to observe for themselves. We see the intelligent laws of God displayed in the countless myriads of snow-flakes in any snow storm, usually all alike, and all perfect. Being ice, yet falling with such fleecy lightness that it requires one foot of recently fallen snow to be equal to one inch of rain. We see different snow storms unlike in the form of flake, yet a large majority of them will be stars of six points, but there are more than one hundred different manners of forming them. Why they have six points instead of three, or five, no one can tell. That subtile substance which we call an electric fluid sometimes tinges those beautiful snow-flakes with light, rendering them distinct in the night, and shedding a faint, though certain, light around. Observation has led me to believe that electricity is in fact a part and parcel of all clouds, being necessary to their existence, and perhaps in a great degree aiding or preventing their formation. If so we must expect that snow clouds also share, and those beautiful tinted clouds of night, sometimes visible, appearing as part of the Aurora Borealis, may be, and I think many times are, the nucleus of coming storms. That the Aurora is in some way connected with our atmosphere, no one now doubts. That the same electricity which electrifies our storms produces it, is equally certain; then why may it not be instrumental in their production. You will seldom see a brilliant display of the Aurora without a storm soon following it. Could we but know a little more we could easily predict storms truthfully; but we must content ourselves with the intellect God has given us, and praise him that we are enabled to turn the lightning stroke from our dwellings, and harness the unknown principle of electricity, causing it to convey our words to the farthest points of earth, without delay. Yet we would fain know many things of our atmosphere, its currents and its aqueous meteors, which time will unfold. Should the next century bring with it as much science as the last, the farmer will not be at a loss today about the weather of tomorrow. Storms of intense thunder and lightning, with hail, will be avoided altogether, and many present wonders of the atmosphere will be understood. And the individual man will be gradually prepared for the nobler contemplation of the works of God, of whose greatness and goodness we can form no adequate conception.

CPSIA information can be obtained
at www.ICGtesting.com
Printed in the USA
BVHW081805061118
532318BV00019B/1467/P